A

MECHANICO-PHYSIOLOGICAL

THEORY OF

ORGANIC EVOLUTION

BY

CARL VON NÄGELI

SUMMARY

CHICAGO

THE OPEN COURT PUBLISHING CO.

LONDON AGENTS
KEGAN PAUL, TRENCH, TRUEBNER & CO.

1898

PREFATORY NOTE.

Mr. V. A. Clark, as a student in horticulture in the University of Vermont, first undertook a critical examination of Nägeli's *Mechanico-Physiological Theory of Evolution* as a part of his regular junior work. After a half year's study and the preparation of a short thesis, Mr. Clark had become so far intimate with Nägeli's work as to make it seem best for him to continue the study through his senior year. This study involved extended translations from the text, including Nägeli's *Summary*, which, considering its difficult accessibility to American students, has been chosen for publication. The work has been done chiefly by Mr. Clark, but has all been under my immediate supervision, and I have given the whole matter a final restudy and revision. Those who have had any experience with similar work will know how impossible it is that all mistakes should have been avoided, and it would be a kindness to the translators if readers would point out any defects, in order that they may be corrected.

F. A. WAUGH.

University of Vermont,
    July 1, 1898.

A MECHANICO-PHYSIOLOGICAL THEORY OF ORGANIC EVOLUTION.

SUMMARY.

IN this summary I shall in general pursue a course the reverse of that which my main work follows.* I shall proceed from the primitive, unorganized condition of matter and endeavor to show how organized micellar substance has arisen in it, and how, from this micellar substance, organisms with their manifold properties have arisen. Since such a synthesis of organisms out of known forms of matter and force is still far removed from a conclusion strictly in accord with physical law, the process becomes comprehensible and obvious only by exact knowledge of the discussion that has preceded. Although the synthetic method reveals more clearly the weaknesses of the theory than do analytic investigations, yet I considered it helpful

---

* See Appendix, Translators' Notes.

to make this presentation in order to give a clearer idea of the mechanico-physiological theory, and at the same time to test its worth.

1. FORMATION OF UNORGANIZED BODIES (CRYSTALS).

When separated and promiscuously moving molecules of any substance in solution or in a melted condition pass into the solid form by reason of removal of the causes of separation and motion (warmth or solvent), they arrange themselves into solid masses impermeable to liquids. These minute bodies grow by accretion, and when molecular forces are permitted to act undisturbed, assume the regular outer form and inner structure of crystals. The number of crystals, their size, changes of form and growth, all depend on external conditions.

2. FORMATION OF LIVING ORGANIZED (MICELLAR) BODIES.

Certain organic compounds, among them albumen, are neither soluble, despite their great affinity for water, nor are they fusible, and hence are produced in the micellar form. These compounds are formed in water, where the molecules that arise immediately adjoining each other arrange themselves into incipient crystals, or micellæ. Only such of the molecules as are formed subsequently and come in contact with a micella contribute to its increase in size, while the others, on account of their insolubility, produce new micellæ. For this reason the micellæ remain so small that they are invisible, even with the microscope.

On account of their great affinity for water the micellæ surround themselves with a thick film of it. The attraction of these micellæ for matter of their own kind is felt outside this film. Hence the micellæ with their films unite themselves into solid masses permeated with water, unless other forces overcome attraction and re-establish a micellar solution (as in albumen, glue, gum), where the slightly moving micellæ show a tendency to cling together in chain-like and other aggregations. Very often there are found, especially in albumen, half liquid modifications intermediate in fluidity between the solid masses and the micellar solution.

The internal and external constitution of micellar bodies depends essentially on the size, form and dynamic nature of their micellæ, since these efficients condition the original arrangement of the micellæ and the insertion in proper order of those formed later. External conditions have slight influence on structure, and affect outer form chiefly in so far as they can mechanically hinder free development.

The micellæ of albumen or plasma are susceptible of the greatest diversity of form, size and chemical composition, since they originate from unlike mixtures of various albumen compounds, and besides are mixed with various organic and inorganic substances. For this reason the plasma behaves, both chemically and physically, in many unlike ways, and in consequence of the variable relation of the micellæ to water, the plasma shows all degrees of micellar solution up to quite solid masses.

3. SPONTANEOUS GENERATION. LIFE. GROWTH.

If molecular forces are so combined in an inorganic substratum that spontaneous formation of albumen takes place, then by the combination of the micellæ the primordial plasma masses of spontaneous generation are given. Within these plasma masses the production of albumen goes on more easily under the influence of their molecular forces than in the liquid without. Hence the compounds present in the organic substratum and capable of forming albumen enter preferably into the masses of plasma, and by intussusception of micellæ of albumen, cause growth. Here life exists in its simplest form. (See page 47.)

Spontaneous generation presupposes the origin of plasma-micellæ from molecules, and hence cannot be brought about by solutions of albumens or peptones, since these are micellar solutions. Life presupposes the intussusception of plasma-micellæ; hence it ceases as soon as the arrangement of micellæ is so far disordered by injurious influences that that process of growth becomes impossible.

The resulting organism must be perfectly simple,

a mass of plasma with micellæ as yet unarranged, because any organization without a preceding organizing activity is inconceivable. For this reason known organisms cannot have orginated spontaneously; a kingdom of simpler beings must have preceded them (*Probien*—the sub-organic kingdom).

The growth of the masses of plasma continues as long as the conditions of nutrition are favorable. If these become unfavorable, a resting period (latent life) or partial or total death occurs, according to circumstances (as lack of nutritive material, lowering of temperature, comparative exsiccation). The growth of plants and animals is nothing else than the continuation of the growth begun in the primordial plasma. This growth still continues wherever the primordial plasma exists.

4. PARTIAL DEATH OF THE INDIVIDUAL: REPRODUCTION.

Since the primordial masses of plasma continue to attract nutritive materials indefinitely and apply them to growth, the nutritive materials are used up in one place and another and the substance which is no longer nourished is in great measure disintegrated. A general condition of equilibrium now sets in, in which the viable plasma masses continue to gain just as much in growth as there is dead plasma broken down and changed back into the original nutritive materials.

In the primordial condition this balancing process is irregular and accidental and remains so even later in many of the lowest organisms. Little by little it becomes phylogenetically more regular by individuals attaining to a more definite size and term of life, while only the germs detached from them remain viable. This phenomenon known as reproduction has a double origin.

A. The portions of primordial plasma that grow to a more considerable size as soft, half-liquid masses break up by the mechanical action of external circumstances into smaller portions of indefinite number and size. This typifies irregular and accidental reproduction of the lowest order.

In the offspring of the primordial plasma division becomes gradually more and more regular as a result of the increasing organization of the substance, and especially as a result of the formation of an envelope about it, till finally in the microscopically small masses, which are now called cells, division into two parts always appears, after these masses have grown to perhaps double their original size. After division the two halves separate from each other and represent independent individuals.

In the further course of phylogeny the division of the cells into two parts takes place regularly. But the cells remain united to each other and form multicellular individuals, which increase by cell division and which at times in the lowest stages are divided at regular intervals into smaller individuals, perhaps even at last into single cells, but from which there are periodically given off cells that develop as germ cells into new multicellular individuals.

B. Another phenomenon which appears in the primordial plasma or its immediate offspring is the death of the greater part of the plasma under certain unfavorable conditions of nutrition, while the smaller part continues to be nourished at its expense and in that case remains viable during the dormant period.

In the offspring this phenomenon gradually becomes free cell formation, which takes place before the resting stage or before the death of many unicellular and multicellular organisms, and which forms germ cells from a part of the contents of the parent cells.

The formation of germ cells by cell division (*A*), or by free cell formation (*B*) is reproduction of the organism. The germ cells are the elements in which the life and growth of the parental individual are continued.

5 MORPHOLOGY OF THE IDIOPLASM IN GENERAL.

The larger part of the unarranged, soft and homogenous primordial plasma, which grows by intussusception, becomes watery soma-plasm, with unarranged and easily movable micellæ. The smaller part is converted in the course of phylogeny into idioplasm, in which at certain favorable points the micellæ that are being stored up under the influence of molecular forces arrange themselves into groups by similar orientations, and hence form bodies of less water content and greater solidity. Each body of idioplasm consists originally of only one group of micellæ, which, however, necessarily breaks up with increasing additions into several groups. On account of the dynamic influence of the groups of micellæ upon their own growth, they become in part more distinct and more definitely separated, in part again differentiated by new irregular intussusception. This phylogenetic process is continued indefinitely, by which the combination of forces produces a new configuration, and conversely, by which a new configuration produces a new combination of forces, so that the body of idioplasm merely takes on a continually increasing complexity of configuration by the action of the internal forces—that is, by the molecular forces of the micellæ of the albumen under the influence of which growth proceeds. This constitutes the *automatic perfecting process* or progression of the idioplasm, and entropy of organic matter. (See p. 47.)

The above described phylogenetic perfecting process of the idoplasm, which operates through internal causes, is scarcely affected by differences of nutrition and by climatic conditions influencing nutrition. On the other hand all those external forces which act as stimuli during a long period of time in an unvarying manner have a very noticeable influence on the intussusception of micellæ in the idioplasm and on the molecular processes going on among the micellæ.   The action of stimuli determines the particular structure of the groups of micellæ added under the direction of the perfecting process.   Thus the configuration of the idioplasm becomes continually more and more complex and at the same time assumes a local adaptation corresponding to external conditions.   This constitutes adaptation of the idioplasm.

6. FUNCTION OF THE IDIOPLASM IN GENERAL.

The unarranged micellæ of the albumen of the spontaneously generated plasma are as yet in no way superior to the unorganized condition from which they have arisen, except in this that under the influence of their molecular forces the formation of similar new albumen micellæ follows more easily.   But as by the further action of molecular forces idioplasmic bodies are formed with groups of smilarly oriented micellæ, the molecular forces of these micellæ amount by summation to molar forces and thereby new chemical processes are introduced; plastic products are formed from plasmic and non-plasmic materials, and molar movements are introduced.   And since idioplasmic bodies are formed under the influence of external stimuli, their plastic products always appear with a definite character of adaptation to environment.

Then, as the idioplasmic body becomes continually more complex in the further course of phylogeny, and consists of a greater number of groups of micellæ differing from each other, the organisms become more complex and differentiate into a greater number of parts, because each group of micellæ of the idioplasm produces its specific effect with regard to inner structure, outer form, and function.

7. DETERMINANTS: THEIR ORIGIN AND DISAPPEARANCE.

Since a particular cluster or group of micellæ of the idioplasm produces a particular phenomenon in the organism, the former is designated as the determinant (*Anlage*, see p. 49) of the latter.   Thus the organism must contain at least as many determinants in its idioplasm as there are different phenomena in its inheritable ontogeny; and if new phenomena appear in it, new clusters of micellæ must previously have been introduced into the idioplasm, or the orientation and arrangement of clusters already present must have been changed.   The formation of such a determinant, whether it concerns the perfecting of the organism or its adaptation to environment, always proceeds very slowly, and as a rule has no effect before its completion.   Hence along with perfected determinants the idioplasm always contains growing and incomplete determinants.

If a phylogenetic line comes under the influence
of other external conditions and other external
stimuli than those which have hitherto acted upon
it, a new and corresponding arrangement of the
micellæ appears phylogenetically in the idioplasm.
At the same time the other adaptation determinants remain either undisturbed, or the new determinant is formed at the expense of related determinants which are already present and which may
at last entirely vanish. Hence along with growing
and complete determinants the idioplasm always
contains likewise weakened and vanishing determinants. From the fact that a phylogenetic race is
thrown repeatedly among different external conditions, it may at last unite in its idioplasm a large
number of developing, mature, and vanishing
adaptation determinants. This number is noticeably
increased if in consequence of interbreeding a
fusion of related idioplasms take place.

8 DEFINITE NOTIONS WITH REGARD TO THE MOR-
PHOLOGY OF THE IDIOPLASM.

Since in the phylogenetic development of the
plasma the thicker idioplasm is differentiated from
the more fluid soma-plasm (§ 5), the former has
the tendency by nature to assume a reticular arrangement. The strands of this network consist, in con-
formity with their origin, of parallel rows of micellæ
extending lengthwise. These rows of micellæ are

combined into more or less complex arrangements, so that the cross section of the strand represents the configuration of the idioplasm.＊

Each ontogeny (individual) begins in a minute germ cell, in which a small quantity of idioplasm is contained. In the cell divisions, by which the organism grows, the idioplasm divides into as many parts as there are single cells, while it continually increases in quantity in a corresponding degree. The ontogenetic increase of the idioplasm takes place by length growth of the strands—that is, by intercalation of micellæ in each row of cells of the strands, which thereby grow in length without changing the configuration of the cross section.†

＊ Nägeli makes his idioplasm ramify throughout the organism in unbroken continuity, much like a system of nerves in the higher animals. This idea with Nägeli was purely speculative. It was known that the protoplasm is in connection throughout the organism, but it has been proved more recently that only the somatic protoplasm is thus connected. The part in which the essential nature of the organism is contained is localized in the nucleus and hence might properly be designated as nucleoplasm, as Weismann suggests. If the idioplasm is localized in the nucleus, it cannot be continuous throughout the system, as Nägeli assumes. But this objection applies only to a detail of the theory and does not affect the fundamental conception,—that of a portion of the protoplasm which is differentiated from the rest and represents a definite molecular structure which determines the specific nature of the organism.—*Trans.*

† Hence, according to Nägeli, every cell of the organism has idioplasm of identical structure. This at once suggests the objection, how can the idioplasm, for instance, of a pollen grain be the same as that of a leaf? Identical idioplasms should always produce identical structures. Nageli attempts to explain this difficulty by attributing the different results to different "conditions of tension and movement," i.e , a dynamical difference between the idioplasms of the different parts of the organism. (*Abstammungslehre*, p 58.)

This idea of differences of structure being due to dynamic rather than to material causes plays a considerable part in Nägeli's theory, but is the point on which he speaks with least certainty—in fact with a noticeable hesitation He does not clearly explain the phrase " conditions of tension

Accordingly, each strand of idioplasm contains all the determinants that the particular individual has inherited in the germ cell, and each cell of the organism is idioplasmatically qualified to become the germ cell of a new individual. Whether this qualification may be realized depends upon the nature of the soma-plasm. In the lower plants this power belongs to each individual cell; in the higher plants many cells have lost it; in the animal kingdom it is possessed in general only by cells normally set apart as asexual or sexual reproductive cells.

The continued phylogenetic formation of the threads of idioplasm takes place by growth in the cross section, which contains the sum of all the determinants and changes in general only when new rows of micellæ are intercalated. But the rows of micellæ of the idioplasm join closely to each other, according to their thickness, so that only rarely new rows can enter, and then only at those definite places where the cohesion is less strong and hence is overcome. The cohesion varies irregularly because the configuration of the cross section, conformably to its origin, is never regular; the disruptive tensions are brought about by the unequal growth in

---

and movement," nor does he give a convincing explanation of the known phenomena as results of the action of dynamic influence

Nägeli is not the only one who posits dynamic rather than material differences as to the basis of diversities of structure More recently, Cope has built up a system of evolution founded largely on this idea.—*Trans.*

length of the individual rows of micellæ.   Dynamic
influences have a decisive effect upon cohesion and
disruptive tensions.   The groups of micellæ of the
configuration already obtained exercise these dynamic influences upon each other; and these dynamic
influences can be modified by stimuli from without.

The idioplasm continually alters its configuration
with its growth in successive ontogenies, but comparatively very slowly, so that it makes a minute
advance from the germ of one generation to the
germ of the next.   The summation of these incre-
ments of advance through a whole line of evolution
represents the race history of an organism, since the
latter is connected only by its idioplasm in unbroken
continuity with the micellar beginning of its race.

9. DEFINITE NOTIONS REGARDING THE FUNCTION OF THE IDIOPLASM.

A plasmic substance causes definite chemical and
physical changes only when it is present in a certain
condition of motion.   The peculiar agency which
the idioplasm has in each ontogenetic stage of
development and in each part of the organism
depends on the activity of a definite group of micellæ
in the cross section of the strand or of a complex
of such groups, while this local stimulus controls the
chemical and physical processes by dynamic influence and by transmission of a specific mode of
motion, even to a microscopically small distance.

The effective stimulus in a plasmic substance is dependent on its own nature and the influence which it receives from without. Which group of micellæ in the idioplasm receives the stimulus depends on the configuration, on the preceding stimuli and on the position in the individual organism in which the idioplasm is found. The determinants have arisen one after another during the whole period of evolution from the primordial cell. The configuration of the idioplasm is a character of phylogeny and the determinants in it have by nature the tendency to develop in the order in which they were formed. Further, since the ontogeny begins as a unicellular organism with the formation of a germ cell, that determinant of the idioplasm comes first to development, which has developed in the unicellular ancestor. Just so the succeeding stages of ontogeny depend for the time being on the development of the determinants having their origin in the corresponding stage of phylogeny. Both causes acting together—the phylogenetic configuration of the idioplasm and the successive morphological stages of development of the individual conditioned on it—necessarily result in the ontogeny being the repetition of the phylogeny.

If the whole remaining line of idioplasmic determinants in an ontogeny has reached development, the development of the germ-forming determinants finally follows as well from the configuration of the idioplasm as from the nature of the organism. The individual is capable of reproduction and the new ontogenies begin in the reproductive cells.

10. TRANSMISSION OF IDIOPLASMIC DETERMINANTS IN LOCAL VARIATION AND IN FECUNDATION.

The automatic progressive or perfecting transformation of the idioplasm is probably active in all stages of development, and proceeds regularly in all parts of the organism, because the idioplasm preserves its configuration at all times and places during the ontogeny. External stimuli impign upon the organism usually at a definite point, but they not only effect a local transformation of the idioplasm but also reproduce themselves in a dynamic manner in the entire idioplasm, which is in unbroken connection throughout the whole individual. The idioplasm is thus changed everywhere in the same manner, so that the germ cells that are given off at any point feel and inherit the effects of those local stimuli.

In the formation of the germ cells in sexual reproduction, the idioplasms of both parents must come into contact with each other, whereupon there results either a material union and formation of a mixed idioplasm or perhaps rather a dynamic action; and through these agencies there is produced a remodeled form which is, however, exactly equivalent to the combined idioplasms entering into it.

Fertilization by diosmose of the spermatic substance is impossible.*

In the idioplasm of a germ cell arising from the crossing of unlike individuals the micellar rows of the individual determinants have sometimes an intermediate constitution and produce characteristics in the organism which are intermediate between the characteristics of the parents. Sometimes the micellar rows derived from the father and mother respectively lie side by side unchanged in the idioplasm of the offspring in distinct groupings and may reproduce in the organism their respective characteristics side by side, or only one of them may develop, while the other remains latent.

On account of the union of both idioplasms as the result of fecundation, two sexually mature organisms are the more able to form with each other a viable germ cell, the nearer they are genetically related—that is, the more nearly the male and female idioplasms correspond in their configuration and chemical nature, because in this case the micellar arrangements are best suited to each other, and the idioplasm of the new fertile germ cell receives its most suitable nourishment from the mother. If, however, self-fecundation or the closest in-and-in breeding often yields products of

---

* This assertion is a direct corollary from the structure of the determinants and the idioplasm. If the idioplasm of the fertilizing cell were to pass through the membrane about the ovum by osmosis, its organized structure would be lost.—*Trans.*

less virility and is avoided by nature, this is the
result of injurious influences which make themselves
felt later on.   This is because incompatibilities may
be present in too closely related idioplasms and
these are sources of weakness in unrestricted development.   The more complicated is the idioplasm, the
oftener this occurs, whereas absolute lack of crossing is not detrimental to the simplest (asexual)
organisms.

11. ACTION OF EXTERNAL INFLUENCES. *

The environment provides the organism above
all with force and matter for its life processes.   It
causes no permanent variation and has only an
ontogenetic significance, if the limits of the idioplasmic elasticity are not exceeded; it maintains the
growth and metabolic assimilation of the individual,
and conditions individual (not hereditary) differences, which constitute "nutrition varieties." (See

---

* In order to explain adaptations Nageli assumes that external influences, if acting at the same point in a given manner for a long time, may
induce slight adaptive variations which are perpetuated and increased.
On the important subject of adaptation in general Nageli is almost diametri-
cally opposed to Darwin and Weismann.   Nageli assigns to the principle of
utility a very limited sphere; Weismann regards adaptation as all-powerful.
According to Nágeli, the organic world would have become much what it is,
if natural selection and adaptation had performed no part in the operations
of nature   He aptly says, that natural selection prunes the phylogenetic
tree, but does not cause new branches to grow.   He allows that the principle of selection is well suited to explain the adaptation of organisms to
their environment and the suitableness and physiological peculiarities of
their structure, but he asserts that in the definiteness of variation of plants
and in their progressive differentiation there is evidence of a higher and
controlling perfecting principle.—*Trans*

page 30.) These appear as the direct results of operating causes.

When the stress of environment exceeds the limits of idioplasmic elasticity, its influence brings about permanent variations, which are imperceptibly small, it is true, in the single individual, but which, when the stimulus is active for a long period of time in the same manner, increase to perceptible magnitude. These variations are inheritable in the phylogenetic sense and contribute to the formation of varieties and species; they always appear as the results of more or less secondary reactions which make their appearance with stimuli exerted by external causes

External stimuli exerted on the organism are reproduced in the idioplasm. Since the stimulus is discontinued with each change of the ontogeny and only the idioplasm persists, permanent variations are produced only in the idioplasm by those conditions that produce visible transformations in the mature organism.

The phylogenetic action of external stimuli gives the definite character of adaptation to the idioplasm as it becomes more complex from inner causes and probably these external stimuli have the power to alter this impress only as new idioplasm is automatically formed.

If an external cause acts continuously upon a phylogenetic line, the corresponding variation of the idioplasm reaches, after a time, a maximum, and thus comes to an end, either because the nature of the substance permits no new rearrangement or because the stimulus is no longer active. The cessation of the stimulus results from a micellar rearrangement which indicates the character of the adaptation. If the action of the stimulus lasts for only a short time, the incipient rearrangement of the idioplasm stops, or proceeds independently on account of the impulse received, and the determinant becomes capable of development, even after the impulse has long ceased to act.

Since various intervening transpositions follow upon a stimulus in the organism, the final result which appears as a reaction may turn out variously. The same external causes may, according to the nature of the organism and other circumstances, have very unlike variations as a result. But the internal rearrangement produces in a definite case very definite variations.

On account of the various intermediate steps it is often difficult to discover the external cause of a given adaptive variation. In many cases we recognize it without difficulty in a definite mechanical process or in warmth, light or evaporation. For the most part the stimulus awakens in the organism merely a want, which the reaction of the organism endeavors to supply. Hence it appears that want or lack alone is able to bring about such reactions.

Moreover, in the sphere of sex, electric(?) attractions and repulsions co-operate between the idioplasmic determinants to produce phylogenetic variations.

The adaptations of the fully developed organism, which are the results of external influences, consist either only of a specific molecular character (irritability), by virtue of which the individual is capable of responding to those influences with temporary or permanent phenomena, or they consist of finished arrangements. The latter have, in general, a double function: either they protect the organism from external influences whose results they are, or they place it in a condition to apply such environmental influences to their advantage. The preponderance of the one or the other led to the development of the plant or the animal kingdom. In the one case the primordial plasma formed in the cellulose cell wall a stimulus-proof covering. On account of this cell membrane being insensible to stimuli, adaptations in the plant kingdom were restricted essentially to the spheres of nutrition and reproduction. In the other case the irritability and mobility of the primordial plasma increased so that it was placed in a condition to avoid the irritant or make it serviceable by accommodating itself to it. The cells sensible to irritants led in the animal kingdom to the formation of organs of sense and the nervous system.

12. CONDITIONS OF PHYLOGENETIC DEVELOPMENT OF THE DETERMINANTS.   ATAVISM.

In the primordial condition, formation and development of the determinants coincide, since the plasma constituting the organism possesses the capability of growing by intussusception of new micellæ and of changing this growth through the action of inner and outer causes. But as the primordial plasma differentiates into idioplasm and soma-plasm, the formation of determinants consists in the transformation of the idioplasm, while the development of determinants consists in the production of soma-plasm and of non-plasmic substances under the influence of the idioplasm.

Only the mature determinant is able to develop, especially if, at the same time, a related and heretofore active determinant must be forced back into the latent condition.   But the determinant of an absolutely new form of adaptation, which does not take the place of a preceding one, must develop enough before it can become outwardly manifest, for it to be possessed of a sufficient amount of molecular energy to render its activity possible.   For this reason the characteristics of the developed organism change abruptly, notwithstanding the fact that the transformation of the idioplasm has proceeded very gradually.

The configuration of the idioplasm becomes continually more complex through the automatic action of the perfecting process, and by this means the organism ascends to higher stages of organization. Hence the viable determinants of organization or perfection are always overtaken after a certain time by that movement and forced into the latent condition. They then become continually weaker, and are at last completely destroyed. Only in the first period after their becoming latent can such determinants pass again into a developmental condition and thus allow the organism to revert to the next preceding stage of organization.

Since the configuration of the idioplasm, which becomes more complex from internal causes, always assumes a definite character of adaptation in consequence of the action of external causes, the adaptation determinants capable of development may become more and more weakened and at last latent when other external causes produce other adaptation determinants. But these determinants may be revived by the renewed activity of the former causes, and thus rendered capable of development   Hence· the organism may show the most various reversions with respect to its adaptations   But in such reversions the earlier forms never quite return, because in the meanwhile the idioplasm has changed somewhat in consequence of its automatic progress, and therefore lends to the adaptations which assume the earlier character a somewhat different expression.

13. ONTOGENETIC DEVELOPMENT OF THE DETERMINANTS.

Since the capability of the primordial plasma to grow is the original and only vital quality (*Anlage*), the whole ontogeny in this first stage consists in the growth of the detached parts to the adult size. In the same way the development of the determinants in all the following stages is nothing more than the growth of the substance detached as a germ cell after the manner of the changes in the character of the idioplasm in the course of phylogeny. In this manner all determinants may in the lower stages of organization reach development, but in the higher stages an increasing number of them must remain latent.

Among the viable determinants there are some that develop unconditionally during each ontogenetic period; there are also alternative determinants of which one or the other unconditionally develops; lastly, there are some that develop only under favorable circumstances. Which of two alternative determinants shall develop depends sometimes on internal, sometimes on external causes, according as the specific determinant has arisen phylogenetically through the action of internal or external causes. Climatic and nutritive influences especially affect the appearance of indefinitely developing determinants. Just so, when a determinant may develop repeatedly (as is so common in the plant kingdom) it depends especially on nutrition whether the corresponding phenomenon is repeated at intervals of greater or less length   A weakened determinant is sometimes temporarily developed by the operation of a definite stimulus.

If the integrity of the organism sustains an injury in consequence of abnormal interferences, determinants develop exceptionally at unusual points.   The process is induced by accumulation of nutritive matter and by external stimuli under the force of necessity, to which the injured organism is sensible.

14. ESSENTIAL NATURE OF THE ORGANISM.

The essential nature of a thing is the sum total of its causes and effects.   Organisms arise from a germ cell which consists of idioplasm and in turn they produce like germ cells.   Their nature depends also on their idioplasm, *i.e.*, on the sum total of their idioplasmic determinants.   Observation of organisms, even in their fullest life history, gives us an imperfect and even false conception of their true nature.   This is because observation reveals only the outer gross characters, and even these in a modification dependent upon accidental effects of nutrition, and does not reveal the finer characters founded in molecular physiology and morphology, and especially the characters latent in the idioplasm.

For the examination of idioplasmic differences

we are restricted to visible characters.   Hence a
knowledge of the nature of an organism presupposes a complete investigation of its characters in
their succession during the whole ontogeny.   The
results must. however be tested and completed by
comparison with other organisms and by the most
comprehensive experimental procedure, possible, (as
by culture under various conditions, and crossing
with nearer and more remote relatives).   The characteristics of nutrition varieties and accidental
crosses must be separated from specific characteristics by experimental procedure, and latent determinants must be brought out by the same means.

15. REPRODUCTION, AND RELATION BETWEEN PARENTS AND OFFSPRING.

Reproduction is nothing more than a transition
from one generation to the next following, medi-
ated by the idioplasm of the germ cell.   In asexual
(monogenic) reproduction there is continuity of the
same idioplasm.   Therefore the parent continues in
the offspring its specific life, as the stem continues
its specific life in the branch.   All the peculiarities
conditioned by the idioplasm remain unchanged in
the offspring.   The latter, as the immediate continuation of the preceding ontogeny, starts from the
point at which the germ cell left it, so that immediately after the germ cell is separated at the close of
the ontogeny or before, the offspring passes at one

time rapidly through the whole ontogeny, at another only the remainder or a part of it (the latter in alternation of generations and in asexual propagation of phanerogams).

In sexual (digenic) reproduction the formation of the germ cell is brought about by the union in equal parts of both parental idioplasms. The offspring is the organism resulting from the union of the force and matter of the parents, and represents in its nature the united continuation of their ontogenies. The characteristics of development of the child depend however on the viability of the determinants of the mingled idioplasms in which a new equilibrium has been formed. Hence if the child bears more resemblance to the father or to the mother, it follows that some of the inherited determinants develop while the others remain latent. If the child has certain visible characteristics more marked than either parent, it becomes possible only by the development of determinants which had previously been latent. The fact that the mother furnishes the germ cell with nutritive plasm and that she nourishes it for a considerable time does not increase the number of maternal determinants nor their capability of development.

If two corresponding characters, one derived from the father, the other from the mother, come into conflict in sexual reproduction, the one or the other, or even a third alternative characteristic, which heretofore was present as a latent determinant, may develop in the child. But also both parental characters may appear at once and in various combinations. Whether the development follows in the one way or the other depends on the strength of the individual determinants, on the kind of their idioplasmic arrangement, and on their agreement with the nature of the newly formed idioplasm.

16. HEREDITY AND VARIATION.

If heredity and variation are defined according to the true nature of organisms, they are only apparent opposites. Since idioplasm alone is transmitted from one ontogeny to the next following, the phylogenetic development consists solely in the continual progress of the idioplasm and the whole genealogical tree from the primordial drop of plasma up to the organism of the present day (plant or animal) is, strictly speaking, nothing else than an individual consisting of idioplasm, which at each ontogeny forms a new individual body, corresponding to its advance.

In this idioplasmic individual the *automatic* or *perfecting variation* is always active, so that the idioplasm of a phylogenetic line always grows by propagation of the determinants contained within it, as a tree grows larger through its whole duration of life by branching. On the other hand the *adaptation variation* caused by external stimuli is present only in those periods of the phylogenetic line in which the idioplasm, and together with this the individual, do not possess the obtainable maximum of adaptation to their environment for the time being. Both of these variations of the idioplasm take place so slowly that only after a long series of generations do the new determinants become capable of developing and revealing themselves in the transmutation of visible characters.

Aside from the phylogenetic variations already named, which take place according to the measure of ontogenetic growth, the idioplasm undergoes, as a result of crossing, as well as in changes of the ontogeny, *gamogenic variations* which may be designated as stationary, since in the mingling of sexually different idioplasms there arise only new arrangements of determinants already present, but no new formation of determinants takes place. Hence in this way arise also new combinations of developmental characteristics.

As a result of external injurious influences, abnormal variations, or *pathological variations*, appear in the idioplasm. These consist of disturbances of equilibrium, which take place also without new formation of determinants. Thereby the determinants already present are caused to develop in abnormal relations, and mostly in reversions.

Apart from the inheritable variations of the idioplasm just enumerated, and the transformations of visible characters involved in it, the soma-plasm and the non-plasmic substances experience, by the influence of nutrition and climate, greater or less variations, which constitute *nutrition varieties*, and since the idioplasm remains unaffected in general, last only so long as the causes which called them forth.*

If we have in mind the inner nature of the organism, there is, properly speaking, no such specific phenomenon as heredity, since the phylogenetic line is a continuous idioplasmic individual. In this sense heredity is nothing more than the persistence of organized substance in a movement in which variations are automatically induced, or the necessary transition of one idioplasmic configuration into the next following. It is present, not only among plant and animal individuals which are ontogenetically separated, but also everywhere within these individuals, where individual parts (cells, organs) follow each other in time. Hereditary phenomena are those that necessarily pass over to following generations, and in general those that are located in the idioplasm, since non-idioplasmic substance can be hereditary only through a limited number of cell generations.

---

* Nägeli, like Weismann, arrives at the conclusion that acquired characters are not inherited. He was not content, however, to rest the generalization upon purely speculative grounds, but undertook the experimental demonstration. After seventeen years of work by himself and son, especially upon several species of Hieracium, he satisfied himself that his theory was true to the facts. We all know now how far he fell short of settling the question.—*Trans.*

Variations and heredity are generally estimated, not according to the inner nature of the mature individuals, but according to their relation in successive generations, since heredity is assumed when the ontogenetic characters remain the same, and variation when previously latent characters become visible. But these phenomena belong to another department of science; they concern the possibility and reality of development of the idioplasmic determinants.

17. VARIETY, RACE, MODIFICATION.

From the multifarious variations of organisms proceed various categories of kinship. *Varieties* arise by extremely slow changes in the idioplasm due to the perfecting process and adaptation. Since these are conditioned by the same causes, they follow in all individuals of the same variety in uniform manner. Varieties are uniform, entirely constant under the most various external conditions, in general cross only with difficulty with related varieties, are not varied by accidental crosses, and persist through geological periods. Varieties belong to feral nature rather than to culture; they can assume all possible modifications without injury to their specific characteristics, but can show no distinctions of races, for all beginnings of race formation are destroyed by free intercrossing. They differ from species only in that they are to be designated as more closely related species, or species as more

remotely related varieties.   Every other distinguish-
ing characteristic is wanting.

Races arise from gamogenic or pathological
variations of the idioplasm.   In the former case they
presuppose crossing between related varieties or
species, in the latter case an increased sensibility
and weakening of the idioplasm.   Very often both
causes co-operate, since crossing follows more easily
when the idioplasm is weakened by hurtful influences and since the irritability and weakening of
the idioplasm increases if crossing has preceded.
Race formation begins in single individuals.   Among
several individuals it begins in various directions
because the causes are different and hence may dis-
play a great multiformity.   Races are distinguished
by more or less abnormal characteristics; they arise
quickly—often in a single generation—and present
various degrees of stability.   This stability is insured
to some extent only by the strictest in-and-in breeding.   All races disappear through crossing, likewise
many races that have arisen from pathological
variations disappear even in sexual reproduction
(in self-fecundation).   Races belong exclusively to
cultivation, where they can develop and exist pro-
tected from free intercrossing.

While varieties and races arise by progressional
or stationary variation of the idioplasm, *modifications* are produced by such influences of nutrition
and climate as act only on the soma-plasm and the

non-plasmic substances, and hence do not give rise
to inheritable characters in the organism. Modifications persist only so long as their causes, and
under other environments immediately pass over
into the modifications corresponding to them. The
transition is completed in the lowest plants during
a limited number of cell generations; in an individual of the higher plants on the same stem during the
growth of a single year. Each variety and each
race appears clothed in a definite modification, and
can change it within a range peculiar to itself.*

18. SOCIAL AND INDIVIDUAL ORIGIN OF SPECIES.

The species arises neither from the *nutrition
variety* nor from the *race;* it is always a more
advanced variety, and hence species formation is
identical with variety formation. Cause for variation and consequently for variety formation is always
shown, either when, environment remaining the
same, the automatic variation of the idioplasm has
advanced so far that the ontogeny is raised to a
higher grade of organization and division of labor,
or when external stimuli act for a sufficiently long
time in a manner not in harmony with the previous
adaptation. Hence various varieties arise easily
from a uniform kinship, when these are thrown
among unlike external influences by local separation, because in the separated places on the one

---

* The distinctions which Nägeli here erects are, of course, purely arbi-
trary, and his definitions are suitable for use only in his own thesis.— *Trans.*

hand the automatic evolution proceeds with unequal rapidity, and on the other hand adaptation takes place unequally.

But in general different varieties arise socially from a uniform kinship. This is because the related individuals living together are unequally stimulated on account of the great inequality of external influences which may exist at the smallest distances; and also because with slight individual differences unlike reactions often follow upon the same external influences. If identically similar individuals are equally inclined to very different reactions toward the same stimulus, sometimes the direction of the first variation decides the character of the adaptation and therefore the nature of the variety, because the variation, when once begun, progresses unswervingly even under somewhat different circumstances.* Hence divergent variations are found growing together in all places, which variations have begun at different though neighboring points by transformation of the idioplasm and are soon intermingled on account of the easy dissemination of seed.

The social formation of varieties is not in general interrupted by crossing, a process which governs only the formation of races. It is confirmed according to experience by the universally recurring fact that several beginnings of the most closely related

---

* It is interesting to compare this statement with Weismann's recent theory of Germinal Selection.— *Trans.*

varieties appear together not only in the same region, but even at the same points, while the geographical distribution of the more marked varieties and of related species offers no conclusion as to their origin, but only as to the last great migration of the plant world, because they arose before this period, as indeed appears from their distribution.

Just as different varieties arise simultaneously from one kinship at the same place, the same variety may arise in places far separated, when the analogous external exciting causes occasion an identical transformation in the idioplasm. The experimental proof lies in the fact that like beginnings of varieties often appear at great distances from each other.

An apparent social origin of varieties is indicated, when, after having come together in migration, they first develop the unlike determinants which they have gained in various locations. An apparently individual origin of the same or different varieties is indicated, when the formation of the determinants take place at one and the same place, but their development follows only after the kindred has been scattered by migration.

19. GENERAL RELATION OF THE PHYLOGENETIC LINES IN THE ORGANIC KINGDOMS.

Since the nature of an organism is contained in the sum of its idioplasmic determinants alone, the evolution of a phylogeny consists in the evolution of the idioplasm. This is perceived from the succession of the visible ontogenetic characteristics which in general run parallel with it. The idioplasm varies in two ways: (1) by an *automatic perfecting process;* (2) by *adaptation to environment.*

By virtue of the *automatic variation* of the idioplasm the ontogenies of a phylogenetic line attain to a continually more complex organization and greater differentiation of function. In this differentiation, however, only the qualitative differences are of importance; quantitative and numerical gradations may be disregarded. The more complex admits of more combinations than the simpler; hence if a phylogeny reaches a higher stage by automatic evolution it may branch into several lines, of which each appears as the continuation of the parent stock.

Since *adaptive variations* depend only on the transmutations of environment, an organism may rise to a higher organization and division of labor by continually adapting itself to the changed environment. But the organism may also change its adaptation while it remains at the same stage of organization. And since the adaptive variation is quickly perfected as compared with automatic evolution, although extremely slowly as compared with the duration of the ontogeny, an organization may change its adaptation several times while it remains at the same grade of organization and division of labor.   Since there are also numerous different kinds of adaptation, a phyletic line may divide at each point into several adaptive forms, which appear in the taxonomic system as species, genera, often even as whole families, while in other cases various degrees of organization have appeared in one family.

20. LAWS OF EVOLUTION OF THE PLANT KINGDOM.

In the sub-organic kingdom, which precedes the plant and animal kingdoms, (see page 5), there are gradually formed from the spontaneously generated plasma independent cells with their characteristic properties, *i.e.*, growth by intussusception of micellæ, formation of a plasmic cuticle, and a non-plasmic membrane about the same, division of the cells, separation of the cells thus formed, and free cell formation within the cell contents.   These properties are inherited from the sub-organic kingdom by the plants and animals which follow in the next stage of phylogeny.   The evolution of the plant kingdom proceeds through the following regular processes, which continue to operate through the entire phylogenetic series.

Law of Phylogenetic Combination.—The simplest of all plants are cells of round form, which grow and reproduce themselves by division, budding or free cell formation.   From the fact that the younger generation of cells, instead of separating from each other and growing to independent plant individuals,

remain united with each other, multicellular plants
arise from unicellular. The same transformation
of the reproductive cells into non-separable tissue
cells is repeated several times in multicellular plants
and serves to enlarge the individual. There is
manifested in this phylogenetic process the tendency of the plant to combine in the higher stages
into one complex whole those parts which in the
lower stages tend to be independent. A similar
unifying tendency is revealed also in those plant
members which have arisen by differentiation and
represent a system only by their being connected at
certain points. These combine in the higher stages
and form ultimately continuous tissues.

*Law of Phylogenetic Complication or Ampliation,
Differentiation and Reduction.*—The cells, and, in
general, the parts of plants which lie near each other
in space or follow upon each other in time, are
always alike in the lower stages. By differentiation they become unlike, so that the sum of
the functions which at first fall to the lot of
all parts without distinction now is shared among
the individual parts. By this means each part can
perform its own special function so much the better.
Differentiation is repeated in the course of the phylogeny, since at first all parts of an ontogeny diverge
into two or more parts, then the parts of these parts
divide again, etc. Along with this process of division another process is always active, which, as it

were, prepares the way for the former, namely, ampliation, by virtue of which the growth of the whole ontogeny or of single stages of it undergoes a quantitative increase, so that an organ acquires a greater number of cells, and an individual a greater number of organs.   After this increase in number of parts in a stage of ontogeny, differentiation follows as far as the nature of the functions permits, by the parts most separated passing into each other by intermediate gradations.   By the further phylogenetic process of reduction the intermediate forms are suppressed.   At last only the extreme products of differentiation lie near each other in space or follow upon each other in time; and these products are as limited in quantity and number as possible.

Along with the above named phylogenetic processes, which take place by the automatic increase of the idioplasm, external influences are always active.   These lend to the organism at times a local stamp corresponding to its environment, and follow the law of adaptation.

21.   ALTERNATION OF GENERATIONS IN RELATION TO PHYLOGENY.

Since the simplest plants are cells and the more complex ones are formed from cells, a whole phylogenetic line may be regarded as a series of cell generations following one after another.   In the lowest forms all cell generations are like each other;

in all others they show differences which become continually greater and more numerous. Thus alternation of generations in cells exists, because the successive generations become more and more complicated at each succeeding period. Among these periods the ontogenetic period or ontogeny embraces all generations from one cell to the return of the exactly similar kind of cell. In the lowest forms of cell differentiation the cells of successive generations are all independent; the ontogenetic period consists of a cycle of generations of unicellular plants. Later the cell generations of an ontogeny are united by parts into plant individuals; the ontogenetic period consists of a cycle of multicellular and unicellular, or only of multicellular plant generations. If all the cell generations of an ontogenetic period have been united into a single individual, the successive plant generations are alike and alternation of generations has ceased.

The unlikeness of the generations arises either from inner causes of temporary differentiation alone, or by temporary differentiation which receives a definite imprint by the change of seasons. But in the latter case the characteristic of adaptation is again lost in the course of the phylogeny and alter- nation of generations follows then without regard to the season. If the given adaptation is united in the lower plants with alternation of generations during the ontogenetic periods, one of the unlike plant generations is repeated an indefinite number of times (repetitional generation), while the other unlike plant generation appears only once and then at the beginning of the resting stage and remains latent in the form of a resting spore till the beginning of the next period of generation. With this peculiar transition generation, which has arisen in the lower stages asexually, and in the following higher stages by the union of a male and a female cell, and which hence is hermaphrodite, there are generally associated later two other single generations—*viz.*, a generation preceding and one following the hermaphrodite, the former as a sex-producing generation, the other as a sex-produced generation.

The phylogenetic significance of the alternation of generations consists in its representing a transition stage from the unicellular to the simpler multicellular and from the latter to the more complex multicellular plants. The plant generations of any phylogenetic stage increase by ampliation, become unlike by differentiation in time (alternation of generations), and unite in a plant individual, whose unlike ontogenetic stages correspond to the unlike plant generations of the earlier ancestral series.

22. MORPHOLOGY AS THE SCIENCE OF PHYLOGENY

All organic phenomena belong, according to their causes, to two different classes: (1) Those belonging to one group are the results of external influences in each ontogeny and are not inherited; they represent nutrition varieties, are experimentally demonstrable, and constitute the subject matter of experimental · physiology. (2) The others are inherited and again transmitted; they belong to the physiology of the idioplasm. This subject is mainly occupied with the origin of the determinants, hence with the formation of varieties and species. It is not the subject of experiment, and constitutes the phylogeny or the physiology of the formation of determinants. A sub-division of this subject is occupied with the development of the determinants already present, hence with the formation of races. It is elucidated especially by experiments in crossing and may be designated as the physiology of the development of the determinants.

The morphological phenomena which find their application in taxonomy, belong exclusively to phylogeny. Their ontogenetic history does not explain their true significance; this can be known only in a phylogenetic way by comparison of one phenomenon with those phenomena from which it has arisen in the course of evolution.

23. PLANT CLASSIFICATION FROM THE STANDPOINT OF PHYLOGENY.

Spontaneous generation has taken place at all times and in all places, in as far as the necessary conditions were concurrently present. (See page 47).

After spontaneous generation the automatic phylogenetic evolution begins and advances constantly.
Consequently the phylogenetic line rises from time to
time to higher stages of organization and division of
labor, but dies of old age if the automatic perfecting
process ceases.   The phylogenetic lines of organisms now living have therefore an unequal age;
those of the most highly developed plants and animals
had their origin in the earliest periods of organic
life, those of the lowest organisms in the most recent
periods.   Hence no general genetic relation exists
among lines now living; only those that are nearly
related and have reached approximately equal stages
of organization may be regarded as branches of the
same phylogenetic stock.   A phylogenetic plant system does not exist in fact, but only in figure.

If genetic relation between two races is assumed,
either as a reality or as a symbol, the degree of
relationship is determined in a theoretically exact
manner by the number and length of the phylogenetic steps which are found either between them
both or between them and the common starting
point, according as races belong to the same or
collateral lines.   The fact that two organisms belong
to the same line of descent is recognized from the
ontogeny of the higher including the ontogeny of
the lower.

Since only a proportionately small number of
known forms can appear as types of the supposed

stages of evolution, only a few phylogenetic lines, and these only in a general way, may be established, on account of the great incompleteness of the present plant world. Such a line proceeds from the green filamentous algæ through the liverworts to the vascular plants. Among the phanerogams, apparently so numerously represented, only phylogenetic series of individual organs can be ascertained, but no phylogenetic series of families. A phylogenetic system of phanerogams is not to be hazarded in the roughest outline. Even the relative rank of the two chief divisions of the angiosperms, the monocotyledons and dicotyledons, is a matter of question, as also which family in each of these divisions is to be considered the most perfect.

APPENDIX.

TRANSLATORS' NOTES.

The Mechanico-physiological Theory of Evolution,
(*Mechanisch-Physiologische Theorie der Abstammungslehre*), by Carl von Nägeli, was published in
Munich and Leipsic in 1884 in a large octavo vol-
ume of 822 pages, including two large appendices.
The *Abstammungslehre* proper, including the summary, occupies 552 pages, and constitutes, in its way,
one of the most important contributions to theoret-
ical biology. It is difficult to understand how a work
of so much consequence should have received such
comparatively small notice in this country, especially
as Nägeli's theories seemed calculated by nature to
appeal much more strongly to American students
than do, for instance, those of Weismann, who has
been studied ten times as much as Nägeli. This is
doubtless due, in part, to the fact that we have had
no English translation of Nägeli's work, a circum-
stance much to be regretted.

The foregoing translation of the summary from
Abstammungslehre goes but a small way toward
making Nägeli's theories accessible to English-read-
ing students, but it will, at least, be better than
nothing. The work covers a great range of sub-
jects, all, however, having a certain relationship to
each other. In the main part of the book the discussion is presented in the following order. (1)

Idioplasm as bearer of the inheritable determinants; (2) Spontaneous generation; (3) Causes of variation; (4) Determinants and visible characters, in which the origin and function of the determinants is presented; (5) Variety, race, "nutrition variety," heredity and variation; (6) Criticism of the Darwinian theory of natural selection, in which the author urges seven objections to that theory; (7) Laws of evolution of the plant kingdom; (8) Alternation of generations from the standpoint of phylogeny; (9) Morphology and classification as phylogenetic sciences; (10) A comprehensive summary of the whole work, a translation of which is given in the foregoing pages.

In the first part of the work Nägeli sets forth his micellar theory of the structure of organized bodies. This is one of his most important contributions to science. Until recent years it has been the only theory given in botanical text-books. At the present time its only competitor is Strasburger's lamellar theory, and even this has not superseded Nägeli's work to any great degree.

The reader who may not be familiar with the micellar theory will find the general idea from the following brief sketch adapted from Vines's *Plant Physiology:*

"Nägeli's micellar theory was developed from his study of organized bodies, especially of cell walls and starch grains. From the behavior of organized substance toward water absorbed by it, he concluded that water does not penetrate into the micellæ, but only among them, thus merely separating them more from each other. He reasoned that if water should penetrate into the micella, its structure would be disintegrated. Hence he argued that organized bodies consist of solid micellæ, which, with their respective films of water, are held together by: (1) The attraction of the micellæ for each other, which varies inversely as the square of the distance. (2) The attraction of the micellæ for water, which varies inversely as some higher power of the distance. (3) The force which holds together the ultimate chemical molecules of which each micella consists.

"Since the swelling up of organized bodies does not take place equally in all three dimensions of space, and on account of their double refraction, Nägeli inferred that in form the micellæ are crystals, probably parallelopipedal, with rectangular or rhomboidal bases."

The law that "bodies attract each other with a force which varies inversely as the square of the distance," has been proven only in its application to the heavenly bodies. Nägeli has applied this law to molecules, unsupported, however, by any evidence other than that of analogy. On the other hand, there is evidence that molecules do not invariably act according to this law.

Spontaneous generation (p. 4) was an important item in Nägeli's doctrine, and might almost be said to be fundamental to it, although it is not really necessary to the internal perfecting principle, which may be regarded as the chief feature of the Mechanico-Physiological Theory. Up to 1865 Nägeli believed in the spontaneous origin of many fungi, and thought that it could be demonstrated. He was obliged to abandon the experimental evidence, but to the close of his life held the views of abiogenesis presented in the accompanying translation.

The characteristic and most interesting feature of the Mechanico-Physiological Theory is certainly Nägeli's conception of an automatic perfecting principle (*Autonome Vervollkommnung*). This conception may be briefly ontlined as follows:

1. The essential part of the reproductive plasm, termed idioplasm, since it divides and passes over from generation to generation, in higher as well as in lower organisms, has a continuous or "immortal" existence.*

2. During this continuous life the idioplasm goes through a development of its own, just as an individual organism goes through a certain cycle of development during its individual life. This development consists in a constantly increasing complexity of structure and differentiation of function.

3. This development is automatic, resulting from internal forces or movements, (*Vervollkommnungs-bewegungen*).

4. As a result of the increasing complexity of structure in the idioplasm the entire organism, which in each generation rearises therefrom, becomes, from generation to generation, more and more complex with greater and greater differentiation of function. Thus the progression of the idioplasm controls the phylogeny of the race. It marks out the course of evolution.

5. Since, according to Nägeli, new life with new idioplasms, may arise wherever and whenever the necessary conditions combine, the present organic world is not made up from branchings of a single original idioplasm, but each race or group may have its own specific idioplasm; and, since this has its own characteristic structure and its own specific internal perfecting forces, it passes through its own

---

*Nägeli's idioplasm corresponds in many respects, though by no means in all, to Weismann's germ-plasm Weismann's idea of continuity or "immortality," which has been so widely noticed, is set forth with equal clearness, though with less emphasis, by Nägeli

peculiar evolution, carrying with it its own de-
pending race of organisms.

The fact that animals and plants at the present
time show such various degrees of organization is
also accounted for on the last supposition, for those
of lowlier organization are merely of more recent
origin and have not progressed so far in idioplasmic
development.

This automatic perfecting principle has been the
mark of much criticism. Some have confounded it
with the mystical *nisus formativus*, or formative
principle of preceding theorists. But, as Weismann
remarks, Nägeli's phyletic force is conceived as a
thoroughly scientific mechanical principle. Nägeli
has simply made application in the organic world
of the principle of entropy, as stated in the mechanical theory of heat. Nägeli himself also compares
his internal perfecting principle to mechanical
inertia. He says, "the force of evolution once
started in a given direction, tends to continue in the
same direction. This constitutes the law of inertia
in the organic world."

Two other matters remain to be noticed. The
first of these is Nägeli's use of the German word
Anlage. We have been unable to give a perfectly
satisfactory translation of this word in its technical meaning. We have received some comfort,
though but little help, from the experience of the
translators of similar works. Selmar Schoenland,
in translating from Weismann, renders it variously
as "germ,' "germ of structure," "germ (of Nägeli)," "germ of Nägeli," "Nägeli's preformed germ
of structure," "preformed germs," "tendency."

Another translator renders the word as "constitutional element." The translation, "determinant," which we have selected is an appropriation of an analogous but not absolutely identical technical term from Weismann's *Germinal Selection*. The use of the word in this connection is open to the objection that it has previously been used technically for a somewhat different idea by another author. M. C. Potter, in his translation of Warming's *Systematic Botany*, following Dr. E. L. Mark, renders the word *Anlage* as "fundament." Dr. H. C. Porter, in his translation of the *Bonn Text-Book of Botany*, renders the same word as "rudiment."

In general the word Anlage means beginning, plan, disposition to anything, and hence involves the ideas of origin, organization and tendency. Sanders defines the word in one of its meanings as: "The act of planning or beginning anything; the act of laying the foundation of any work intended to be carried on toward completion, in order that from the beginning made, a definite thing may be developed or may develop itself"; (*i.e.*, to determine, in the sense of limiting to a particular purpose or direction, hence determinant). "Also, the thing begun or planned, considered as the basis and germ of the further development of that which has already originated."

In its restricted use as applied to organisms it would mean "germ," in the sense of embryonic starting point. More specifically, it is a portion of plastic, organized substance, functioning as an individual and containing potentially an elemental organ plus a formative power. In Nägeli's own words, "There exists an essential difference between the-substance of a mature organism which does not possess the capability of further development, and the substance of an egg, which does possess this capability. By virtue of this difference the egg-substance is characterized as the *Anlage*, or germ of the mature organism. All characteristics of the adult condition are potentially contained in the ovum."

Nägeli was not the first to assume the existence of a unit of organization intermediate between the molecule and the cell. E. B. Wilson, in his *The Cell in Inheritance and Development*, states the case as follows:

"That the cell consists of more elementary units of organization, is indicated by *a priori* evidence so cogent as to have driven many of the foremost leaders of biological thought into the belief that such units must exist, whether or not the microscope reveals them to view The modern conception of ultra-cellular units, ranking between the molecule and the cell, was first definitely suggested by Brucke in 1861.

"This idea of ultra-cellular units is common to most morphologists and physiologists. We are compelled by the most stringent evidence to admit that the ultimate basis of living matter is not a single chemical substance, but a mixture of many substances that are self-perpetuating without their loss of specific character."*

Nägeli's *Laws of Evolution* are also worth special notice. As stated in the body of *Abstammungslehre* they are as follows:

1. Asexual reproductive cells which arise by division, remain united and become tissue cells.

---

* For a fuller discussion of the notion of these hypothetical units of organic existence, see Weismann's Germinal Selection, (Open Court Publishing Co., Chicago, 1896), especially the foot note. page 230.

2. Asexual reproductive cells which arise by budding, instead of separating, become cell branches or branched cell threads.

3. Reproductive cells which arise by free cell formation become bodies which form a part of the cell contents.

4. Parts of a plant which arise by differentiation lie side by side and form a body of web-like or tissue-like structure.

5. A definite and previously limited growth continues, or a definite formation of parts of an ontogeny which has previously been present but once, is repeated   (Ampliation.)

6. The parts of an ontogeny become dissimilar, since the functions which were previously united become differentiated and since new ' dissimilar functions are produced in the various parts.   This differentiation is either one of space between the parts of the ontogeny that appear near each other, or one of time between those that are derived from each other.

7. Parts which have become dissimilar by differentiation undergo a reduction, in which the intermediate forms are suppressed and at last only the qualitatively dissimilar forms with qualitatively dissimilar functions remain.

8. The environment in which plants live operates in different ways, directly as a stimulus or indirectly as a felt necessity and by this means lends to their forms and activities a definite expression of time and place, and thus brings about different adaptations.   These become permanent through heredity, but are again gradually lost if other adaptations supersede them.

Laws 1 to 4 may be expressed as one—the law of combination: Similar parts that are wholly or partly separated have the tendency to unite more and more completely and intimately into one continuous tissue.

The laws of ampliation (5), differentiation (6), and reduction (7), may be summarized in one as follows: While increasing in size the similar parts of an ontogeny become internally dissimilar and the dissimilarity increases as the transition forms of the dissimilar parts vanish. Hence only the extreme forms remain.

It may also interest the reader to know that Nägeli was the first to propose the general theory of cell formation as accepted at the present day.

.

Lightning Source UK Ltd.
Milton Keynes UK
UKHW022028080221
378458UK00003B/518

9 780526 198405